The Meteorite Report

Jean-Baptiste Biot

Relation of a Voyage made to the
Department of Orne
to verify the Reality of a Meteorite
observed at L'Aigle
on the 6th of Floreal of the Year 11
(April 26th, 1803)

Translated, and with an Introduction by
Anne-Marie de Grazia

ALBRECHT DÜRER: Explosion of a Meteorite (the meteorite of Ensisheim, which he witnessed in Basel on November 7th,1492) Painted on the back of *Saint Jerome Penitent* - FITZWILLIAM MUSEUM, Cambridge University

Jean-Baptiste Biot (1774-1862)

The Meteorite Report

RELATION OF A VOYAGE MADE IN THE DEPARTEMENT OF ORNE,

in order to verify the reality of a meteorite observed at L'Aigle on the 6th of Floreal of the year 11 [26th April, 1803]

read to the class of mathematical and physical sciences of the Institut National, on the 29th of Messidor of the year 11 [18 July 1803]

Translated, and with an Introduction by
Anne-Marie de Grazia

Metron Publications

ISBN: 978-1-60377-086-6
LCCN: 2012956238

Princeton, N.J., 08540-1205
USA

STONES FALLING FROM HEAVEN

Reports about "stones falling from heaven" must have existed since earliest mankind. In the late 16th century, Indians of the Chaco in Northern Argentina knew of a large mass of iron which had fallen from the sky in a place which they still called Piguem Nonralta ("Field of the Sky"), which the Spanish colonialists translated aptly as Campo de Cielo. In 1576, a Spanish governor from Paraguay sent an expedition in search of a deposit of iron, which the Indians were said to be using to make weapons. A large mass of metal protruding out of the soil was found, to which the Indians assigned a celestial origin, and samples of which were taken, to be deposited, along with the records of the expedition, in the *Archivo General de Indias* in Seville. The area was investigated again by Spanish explorers two centuries later and the iron deemed to have formed in a volcanic eruption. Samples from these latter expeditions were sent to the Royal Society of London. It is not until the 20th Century that the iron masses of Campo de Cielo were identified as meteoritic - and a huge meteorite it had been, adding up to close to 100 tons of matter, distributed over 26 still identifiable craters. The event, the memory of which had been preserved by the Indians, down to their toponymy, was dated back to some 4,000 to 5,000 years...

Among the earliest meteoritic events having come

down to us from our own Antiquity is the fall of a stone in Phrygia, near the city of Pessinus (120 km SW of today's Ankara), around 470 B.C. This "Phrygian Stone," which was cone-shaped, was worshipped in connexion with the Great Mother Goddess Cybele. During the Punic wars, after an impressive rain of stones over Rome, the Roman Senate consulted the Oracle at Delphi which recommended that the "Mother Stone" of Pessinus be brought to Rome. It was secured as a loan from Attalus 1st, king of Pergamon, and brought in a lavish procession to Rome in 204 B.C., where it was kept on the Palatine Hill and continued to be worshipped under the name of Mater Deum Magna Idaea over the following centuries, as told by Plutarch (ca 46-120 AD). The loan was never returned.

Among reports from Greco-Roman Antiquity, we notice a stone "the size of a cartload" which is reported to have fallen in 467 B.C. in Aegos Potamos in Thrace, and was kept and worshipped there for many centuries. There are reports that a comet was visible in the sky at the time of the fall. It seems that the philosopher Anaxagoras got himself into serious trouble in connection with the event, for describing all heavenly bodies as being mere stones, a sacrilege for which he was beset by an irate mob. Pericles intervened - physically, with his own body - to protect him. Anaxagoras was said, several centuries later, to have predicted the fall of the meteorite, which is unlikely.

The naturalist and statesman Pliny the Elder reports that he was himself the witness of the fall of a meteorite in 70 A.D., when he was procurator in Narbonensian Gaul.

The Chinese historian Ma Duanlin (1245-1325) compiled reports of meteorite falls in China covering 2000 years. A 20th Century analysis of early Chinese reports found 337 well-observed meteorite falls between 700 B.C. and 1920 A.D.

On May 19, 861, a stone of 472g fell, "among thunder and lightning" within the precinct of the Shinto temple of Nogata, on the Japanese island of Kyushu. It has been preserved inside the temple, in a casket bearing the date of impact, which has been corroborated by Carbon-14 dating of the material of the casket.

The Haschar al-Aswad, a black stone masoned into the Kaaba, the holiest shrine of Islam, is said to be a meteoritic stone retrieved from the mountains around Mekka, although its meteoritic nature has not been scientifically ascertained.

The science of meteoritic phenomena in the Middle-East was advanced enough for the Persian physicist and philosopher Avicenna (ca 980-1037) to write that "stones falling from heaven are of two kinds:" some are of the nature of stone, and some are made of iron. It would take almost 800 years for the West to catch up with this simple knowledge.

With the advent of Christianity, the numerous meteorites which had served as objects of worship were banned, discarded and destroyed. Besides identifiying the meteorites as hateful pagan objects, the Church had largely adopted the teachings of Aristotle, for whom the universe was constituted of nestled spheres, of an immutable nature, which revolved in perfect order around the Earth and could contain no stray elements of a foreign constitution or motion. There was a sphere of fire above Earth, separated from it by a sphere of air, and meteorites were thought to be fires from earth ascending to heaven to rejoin with the fire above.

In Europe, the first recorded meteorite of which some remains subsist to these days fell around 1400 in Elbogen, in Bohemia. The exact date is not known.

A true historic sensation was created by the event which occurred at Ensisheim, in Alsace, on November 7th, 1492, just before noon (Columbus had discovered America three weeks earlier): a blazing fire-ball was widely witnessed over Switzerland and Southwest Germany and a 127 kg stone fell in a wheat field outside the town of Ensisheim. The locals rushed to the spot and eagerly started to break off pieces for use as talismans. This was soon forbidden. Maximilian of Habsburg, German King and King of the Romans, not yet Holy Roman Emperor, who was on the war-path against the French King Charles VIII. (in order to retake his fiancée, 13-year old Anne of Bretagne, to whom he had been married by proxy) happened to be in the vicinity, visited the impact site, and wasted no time using the event for political propaganda. He had the meteorite put into chains and hung out of reach in the church of Ensisheim.

Forty kilometers to the South, in the city of Basel, young Albrecht Dürer almost certainly witnessed the fire-ball exploding in the atmosphere and recorded his impression in an extraordinarily vivid, truly "modern" painting on the wooden back of another painting.

Also remarkably, within two weeks, several wood-cuts were made in Basel relating the event, complete with texts by the humanist Sebastian Brandt who was then teaching at the University (his *Ship of Fools* would appear two years later) and were printed as "flying leaves," which were distributed and sold within days all over the German speaking areas of Europe. These two events - Columbus'

Flying sheet about the Ensisheim meteorite by Sebastian Brandt, printed by J. Von Olpe, Basel
(with his motto: Nüt On Ursach - *Nihil Sine Cause*, dated 1492

landing and the invention of the "sensational" press - mark these weeks of 1492 as the true beginning of Modernity.

The Church continued to hold on to the Aristotelian view and the Ensisheim meteorite was categorized as a miracle. Science freed itself only slowly, over the next three centuries, from this double ideological clutch. Newton's conception of the space between the heavenly bodies as being free of all matter, pervaded only by the ether, left no room for small objects, such as stones falling from the sky.

In 1771, Goethe paid a visit to the Ensisheim meteorite and chided peoples' credulity: with all the stones to be found around in the fields, why would anyone need to explain some of them as having fallen from the sky? The French Revolution sequestrated the Ensisheim meteorite as Church property. Scientific consultants accompanying the liberating American armies in 1945 tried to purchase it, but were turned down by municipal authorities. The meteorite is today on display at the town hall.

On September 16th, 1753, a meteorite fell in Luponnas, France, and the event was studied by the astronomer Jérôme-François de Lalande (who had just achieved the first precise calculation of the distance between the Earth and the Moon). He suggested a possibly celestial origin, but his report was met with skepticism.

On September 13, 1768, at 4:30pm, workers in a field observed the fall of a meteorite in the village of Lucé, in Maine, near the road leading to Le Mans. The local priest communicated their story, together with a sample of the stone, to the French Academy of Sciences. In 1777, the sample was submitted to a chemical analysis by the latest methods - the first ever for a meteoritic stone - by the chemists Fougerous de Bondaroy and Cadet de Gassicourt, under the direction of Antoine-Laurent de Lavoisier himself. The results were published in the *Journal de Physique.* The authors concluded that the stone was of terrestrial origin and could have been caused by a stroke of lightning hitting sandstone.

In 1749, in Siberia, between Ababansk and Krasnojarsk, a cossack smith came across a mass of iron of ca 687 kg, of an extraordinary aspect, which was in the possession of a local Tatar tribe. He had it brought to his smithy. Its existence was signalled in 1776 to the German naturalist Peter Simon Pallas (1741-1811) who was on a mission of exploration in Siberia commissioned by Catherine II. Pallas was impressed. The object was made of very large crystals of peridot joined by a matrix of mixed iron and nickel. Nothing resembling it was known. He had it transported to Saint-Petersburg, where it found its way into the prestigious cabinet of curiosities which had been created by Peter the Great. Pallas returned to Berlin, bringing with him a piece of the mysterious object. Meteorites of this very rare type would be called after him *pallasites.* (In a strange twist, another *pallasite,* weighing 198 kg, was found in 1990 near the city of Pallasovka, near Volgograd, a settlement which had been named in honor of Pallas, at a distance of 4000 km from the site of his original find.)

Enters a young German genius of the Enlightenment, Ernst Chladni (1756-1827). Having satisfied his father by becoming a Doctor in Law, he decided to take time off to indulge his talents in music and mathematics and turned to experimental acoustics, which he almost immediately revolutionized through the discovery of the *Chladnische Klangfiguren* ("Chladni Patterns") which are predictable geometric patterns created by sand on a metal sheet set vibrating with the help of a bow or a tuning fork. In 1792, during a visit to his friend, the physicist Georg Christoph Lichtenberg (1742-1799) at the University of Göttingen, he found himself caught up in intense discussions on the subject of fire-balls and meteorites, including mentions of the Pallas stone; Lichtenberg himself had witnessed a fire-ball the year before. Over the following year, as a sideline, Chladni threw himself enthusiastically into the study of all the stories of meteorite falls and of all the so-called meteoritic samples then available.

In 1794, he published the results of his work: *Über den Ursprung der von Pallas gefundenen und anderer ihr ähnlichen Eisenmassen und über einige damit in Verbindung stehende Naturerscheinungen.* ("About the origin of the iron masses discovered by Pallas and of similar ones, and about certain natural spectacles which are connected with them.") He argued against the various explanations hitherto offered by science as to the origin of these phenomena. He strove to establish that the phenomena of fireballs and meteorites were linked. He hypothesized that both originated outside Earth; that fireballs were created by the heat resulting from the friction of bodies entering the atmophere. He suggested that meteorites were small masses of matter moving through space before being captured by Earth's gravity field; that they were either residues of the formation processes of larger celestial bodies, or debris resulting from the destruction of such. Chladni got pretty much everything right...

Yet, Chladni's work and speculations were not

received by the scientific establishment as he had hoped they would. Goethe and von Humboldt dismissed his theses. Lichtenberg himself was skeptical, especially as two of his students, Benzenberg and Brandes, had just carried on measurements proving that meteorites were of purely atmospheric origin; but he conceded that Chladni might still be proven right. The French Academy, of whose opinion Chladni was most hopeful, derided him. One member is said to have worried that one of those stones Chladni wrote about may have landed on his head! He found a more open mind in Sir Joseph Banks (1743-1820), President of the Royal Society in London.

That same year - 1794 - on June 16th, around 7:00pm, a dark cloud approached Siena, in Italy, broke out in a display of red lightning, followed by a tremendous explosion, after which some 200 stones, some weighing 1.5 kg, rained onto the outskirts of the city. English tourists collected some samples. Italian scientists published a study by year's end. William Thomson, of England, published the first mineral description of one of these stones, and stressed the fact that it differed from any known rock of earthly origin.

On December 13, 1795, a field worker witnessed the fall of a 25kg stone near Wold Cottage, in Yorkshire, on the land of Major Edward Topham, a newspaper owner, who saw to it that the event was well covered. The stone was exhibited at Piccadilly. Like the stones of Siena, it was covered with a black crust. Sir Joseph Banks asked a young chemist, Edward Charles Howard (1774-1816) to undertake some analyses, together with a French émigré mineralogist, Count Jacques-Louis de Bournon (1751-1825). He added to the collection a few more samples: one from Campo de Cielo (see above);

one from Agram (i.e. Zagreb) in Croatia, which had fallen in 1751; one which had fallen in Benares, India, in 1798, as well as a sample from the Pallas object. A remarkable similitude was found by the chemists to exist between the "stone-like" meteorites found on sites extremely distant from each other, as far as their mineralogy, chemical composition and texture were concerned, while they displayed at the same time a great difference with all other known terrestrial rocks. Most significantly, a high quantity of nickel was found associated with iron both in the case of the "grains" of magnetised metal contained in the "stone-like" meteorites, and in the iron making up the iron meteorites. In 1802, Howard and de Bournon presented their report to the Royal Society and in 1803, the year 11 of the French Republic, to the French Academy, where the French chemist Nicolas Louis Vauquelin also presented the results of his own, concording, analyses of samples from an other meteoritic event, a stone which had fallen at Barbotan, in the Pyrénées.

Then, on a bright, serene day of the 6th Floreal of the same year 11 (26. April, 1803), at high noon, a fireball raced through the sky over Normandy and exploded above the town of L'Aigle...

Citizen Marais, of L'Aigle, wrote about it to a friend in Paris, Citizen Château, on May 3rd. Citizen Château gave the letter to a neighbor to read, Citizen Charles Lambotin, a student in mineralogy, asking him to alert the Academy of Sciences, which he did, most efficiently, as the letter was published in the issue of 19th Prairial (May 9th!) of the *Journal des Sciences Mathématiques et Physiques de l'Institut National.* Citizen Lambotin went to L'Aigle and collected numerous fragments, which he sold to collectors in Paris.

Marais' letter started thus: "There has just happened in our area a rather surprising miracle..." Bonaparte, possibly alerted by the *Journal,* asked the Minister of the Interior, the chemist Jean-Antoine Chaptal to clarify the matter urgently, especially in view of the spectacular progress that was being made in England at the time - witness: the conference of Howard and de Bournon. Chaptal sent one brilliant subject, Laplace's protégé, Jean-Baptiste Biot, age 29 yet already titulary of the chair for mathematical physics at the French Academy of Sciences, to investigate on the spot...

Jean-Baptiste Biot delivers. His report is unimpeachable. He proves the fact that meteoritic stones fall from the sky and that they are of non-terrestrial origin. The news go around quickly in Europe and everyone clamors for a piece of the L'Aigle meteorite. There will be no looking back.

It will take a little while longer for the new idea to cross the Atlantic, though, witness the following story taken from:

http://www.meteorlab.com/METEORLAB2001dev/metics.htm]

In the cold chill of a December morning in 1807, Judge Wheeler walked from his home in Weston, Connecticut, USA, and was surprised to see a ball of fire moving across the northern horizon. He watched as it passed to a point almost overhead where it flashed several times and disappeared.

A few moments later he heard a great noise. Thunderous and roaring, the noise grew to a frightening level. He then heard the whizzing sound of something falling. As the judge looked up, he observed a small stone strike a nearby building, bounce off, and roll onto the grass. The judge decided to contact nearby Yale University and ask that the event be investigated.

Two very skeptical professors came out to look into the matter, fully prepared to dispel the story of stones falling from the sky. The two professors conducted a lengthy investigation. They knew these stones were different from any they had ever seen and they witnessed local townspeople extracting them from holes in yards and nearby fields. Finally, the two wise professors from Yale concluded the stones must have fallen from the sky.

Eventually the story found its way to the White House in Washington, D.C. President Thomas Jefferson was a scientist as well as a statesman. When he heard this peculiar story he declared it could not be true, but his advisors insisted that the stones were observed falling from the sky and that two Yale professors investigating the incident vouched for its truth.

Thomas Jefferson, President of the United states, responded with great skepticism: "Gentlemen, I would rather believe that two Yankee professors would lie than believe that stones fall from heaven."

A.M. de G.

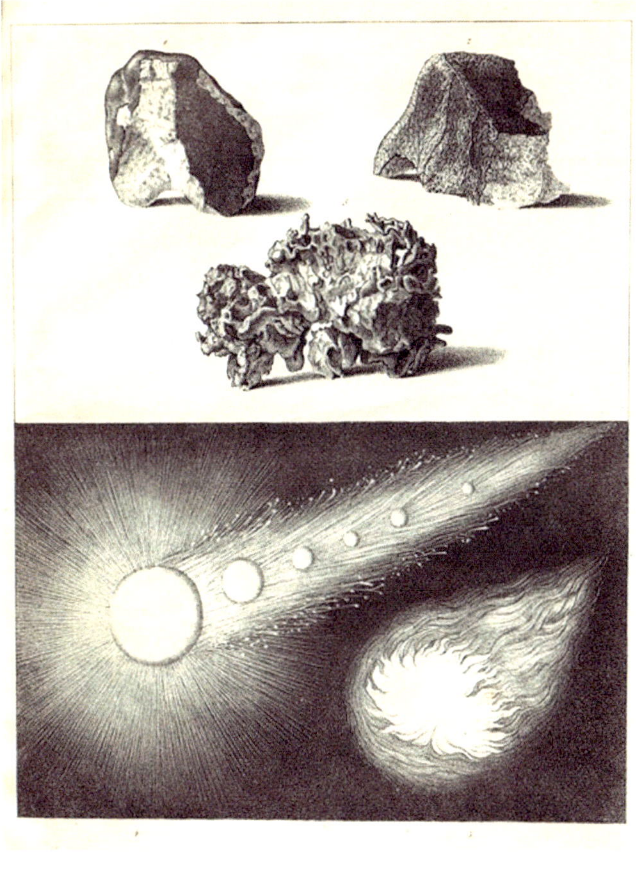

This illustration of a 1838 monograph by Ferdinand Keller: *An die Züricher Jugend auf das Jahr 1838* (Zürich) shows at the top, left: a piece of the meteorite of L'Aigle of 1803, top right a piece of a meteorite fallen at Stannern, in Moravia,, on May 22, 1808, and top middle a piece of the meteorite shown to Pallas in 1771. The picture below represents an enormous fire-ball which appeared on July 3rd, 1762 over Saxony and Brandenburg, before exploding loudly over Potsdam. No meteoritic masses were recovered from this incident.

Jean-Baptiste Biot (1774-1862)

RELATION OF A VOYAGE MADE IN THE DEPARTMENT OF ORNE,

in order to verify the reality of a meteorite observed at L'Aigle on the 6th of Floreal of the year 11 [26th April, 1803]

read to the class of mathematical and physical sciences of the Institut National, on the 29th of Messidor of the year 11 [18 July 1803].

(Translated by Anne-Marie de Grazia)

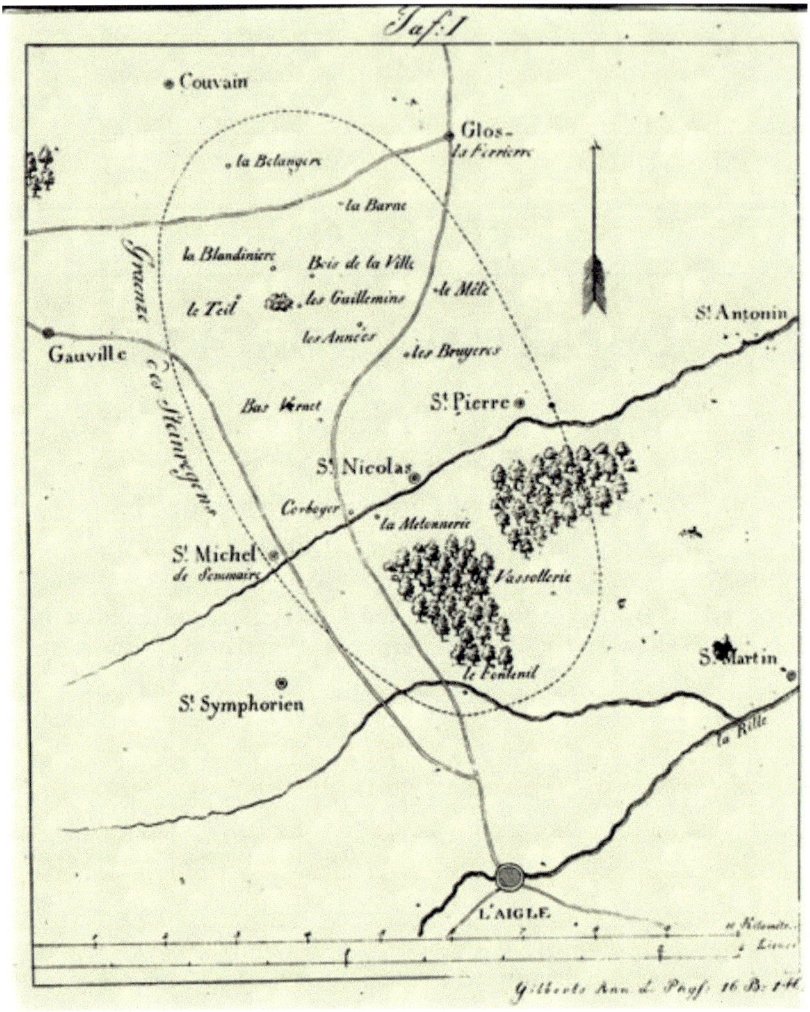

Map of the site of the meteorite fall of L'Aigle made by J.-B. Biot and attached to his report. The delineated ellipse shows the area of scattering of the debris, the heavier ones having fallen at the Southern end and the lighter ones at the Northern end.

THE MINISTER of the Interior[1] having asked me to travel to the département of Orne to gather exact information about the meteorite which appeared in the vicinity of L'Aigle on 6th Floreal of this year, I hastened to fulfill his purpose and I shall give account to this class of the observations which I have collected. May the importance of the subject excuse the multiplicity of details I shall enter into.

Ever since the attention of scientists has moved towards the study of mineral masses which are said to have fallen from the atmosphere, all the resources of critical examination and of experience have been employed to record this astonishing phenomenon and to throw some light as to its cause. In the same time that chemical analysis determined the elements in these masses, distinguished them from the natural products already known, and discovered in their perfect identity the proof, or at least the great probability of a common origin, all the accounts which could have had some relation to same fact have been collected; one has consulted the accounts of the ancients, whose authority has been all too often met with suspicion, and who more and more are being recognized to be faithful witnesses, the more often that occasions present themselves to verify their observations. In order to complete these research and to make its full importance felt, ingenious hypotheses have been brought forth in order to satisfy, in accordance with the laws of physics, to the hitherto observed

[1] The chemist Jean-Antoine Chaptal (1754-1832).

phenomena. Finally, scientists of all classes, of all countries, have united their efforts upon this great question, guided, not by jealous rivalry, but by the noble love of truth.

Without doubt, such unanimous concourse will be remarked on in the history of the sciences. It is both result and proof of their progress. It is a big step made in the study of nature to be able to examine a phenomenon no full explanation of which can yet be entertained, and a courage of this sort can belong only to the most enlightened of men. We must therefore thank our colleague Pictet, who set us a first exemple in the matter at hand by communicating to us the investigations of the English chemists; which investigations a precipitous opinion might have treated as fanciful, but which was discussed in the present class with the reserved enthusiasm thanks to which one equally avoids passing up new truths, and welcoming errors. Indeed, of what help can prejudices be, when everything is lacking to form oneself an opinion? Always, when faced with doubtful questions, the ignorant believes, the half-learned decides, the educated man examines: he does not have the temerity to pretend to put limits to the power of nature. Let us therefore follow zealously, and without letting anything hold us up, the phenomenon which presently occupies us; and if we should finally succeed, as I hope we will, to establish its reality without a doubt, let us not forget that it has been the conceit of wanting to explain everything which has caused it to be rejected for so long.

Of all the probabilities collected up to now about the fall of meteoritic masses, the strongest result from the agreement existing between the identity of their composition, and the identity of the origin which testimonies grant them exclusively. This agreement, already verified in a great number of observations, gives to this probability a value which is very close to a certitude and which is by no means infirmed by the objection that the witnesses are but little educated; for, precisely because of the paucity of their education their testimonies should, if the facts were false, evoke diverse substances, dissimilar circumstances; and for a topic of this nature, in which particular personal interest enters for nothing, the chance of the convergence of testimonies is unique, whereas that of their divergence is infinitely multiple.

Still, it remained much to be desired that such a phenomenon be at least once recorded in an irrecusable manner, and that all its particularities be faithfully collected, as much in order to establish a moral certitude as to its existence, as to know exactly the circumstances which signal it, and which are also necessary in order, if at all possible, for us to trace our way back to its causes, or at least, to prevent us from getting lost in error when searching for the latter.

Convinced of such truth, I felt that only the most scrupulous exactitude and faithfulness could make useful to science the mission with which I had been entrusted. I constituted myself as a witness estranged from all systematic thinking; and, in order to not offer anything which might endanger the trust in that which I am going to report, I will limit myself in this memoir to exposing the facts such as I have collected them, gathered them, and only developing the immediate consequences resulting from their relationships. I will even abstain from examining how they draw close to or move far from hitherto imagined hypotheses.

Before beginning my research, I found it necessary to methodically classify the facts on which I was to principally direct my observations; I therefore put them together in the following table:

- ARGUMENTS
  - PHYSICAL, derived from:
    - The presence of meteoritic stones in the hands of inhabitants of the area.
    - Traces or debris left or occasioned by the meteorite.
    - Mineralogical and geological circumstances of the area.
  - MORAL, derived from:
    - The testimonies of persons who have seen and heard the meteorite;
    - The testimonies of persons who have heard the meteorite but have not seen it;
    - The testimonies of persons who, findthemselves on the spot, have searched for and collected informations about the meteorite and its effects.

Before leaving, I gathered on these diverse questions all the information I could put my hands on. I begged Citizen Haüy to be so kind as to enlighten me as to the mineralogy of the land I was going to survey. Citizen Coquebert Montbret, a correspondant of this class, furnished me with the necessary knowledge about the physical geography of said land. Finally,

Citizen Fourcroy was kind enough to give me a copy of the letters which he had received from L'Aigle relative to the apparition of the meteorite.

I left Paris on the 7th Messidor (June 26th), taking with me a compass, a Cassini map, and a sample of the meteoritic stone of Barbotan,[2] which had been handed there to our colleague Cuvier: I intended to use it as a term of comparison, and I was curious to find out what origin the inhabitants of the cantons where it was said that similar ones had fallen, would assign to it.

But I did not go straight away to the very spot. If the explosion of the meteorite had indeed been as violent as we have been told, its noise must have been heard very far away. It was therefore in conformity with the rules of critical investigation to start by gathering information in distant places about this extraordinary noise, about the day and the hour it had been heard, to follow its direction and to let myself be led by testimonies right to the spot where the meteorite was said to have exploded. In this way, I would be to gather, over a large expanse of land, comparable informations; for, as to the noise, and the very circumstances of the explosion, testimonies would have to concord, wherever they were gathered. Anyway, all stories relative to meteoritic masses have their apparition preceded by a ball of fire. It was important to make out if the meteorite of L'Aigle had been accompanied by the same circumstances, and it was only far from the spot of the explosion that I could ascertain this.

Guided by these considerations, I first went to Alençon, chef-lieu of the Orne département, situated at 15 leagues (1

---

[2] Now a part of Cazaubon, commune of the département of Gers (Midi-Pyrénées). A 6.4 K stone fell there in 1790.

metric league = 4 km; 15 x 4 = 60km)[3] to the south-west of the city of L'Aigle.

On my way, the courier from Brest to Paris told me that on Tuesday, 6th Floreal,[4] at 9 leagues (36km) beyond Alençon, between Saint-Rieux and Pré-en-Pail, he saw in the sky a globe of fire which appeared in a serene sky, on the side of Mortagne, and seemed to be falling towards the north. Some instants later, a great noise was heard, like of thunder, or of a carriage rolling continuously over cobblestones. This noise lasted several minutes, and could be heard despite the noise of the mail-coach which was then driving over a dirt road. The time was three quarters after noon, and the courier told me that he had immediately checked his watch, as the sight had most astonished him. He added that, when arriving in Alençon, he had told this fact in the house where he had spent the night; and this has been confirmed to me since. From the progess of this fiery globe, from the noise, and especially from the hour, I judged that this was the beginning of the meteorite of L'Aigle.

In Alençon, they had vaguely heard about this phenomenon, but they had seen nothing; and no extraordinary

---

[3] The metric system, called MKpS for "mètre - kilogramme - poids - seconde," one of the lasting contributions of the French Revolution, had been adopted by decree of the Convention on the 18th Germinal of the year 3 (7th April 1795).

[4] The Republican Calendar, with its poetic-bucolic names of months, was introduced by decree of the Convention on 14th Vendémiaire of the year 2 (October 5, 1793), in order to wean the nation of the patterns of religious time references. It was abrogated by Napoleon in 1805. It can be seen here htat its division of the month in three "décades" of ten days each did not catch on: the former days of the week continued to be used. This "Tuesday" should have been a *"Sextidi..."*

noise had made itself heard: which is not surprising in a large town, amidst the tumult of a market day. The prefect, the chief engineer of bridges and roads, the professors at the Central School[5] had no knowledge of the meteorite. But if these citizens were not able to give me any direct information on the subject, they furnished me with others, no less useful, by letting me inspect their collections. Citizen Barthélemy, engineer in chief, a man as distinguished by his knowledge as esteemed in the area for his character, has busied himself for five years collecting samples of all mineral substances found in the département of Orne, with the purpose of finding material necessary to industry or for civil construction. In this collection which I examined, nothing resembles meteoritic masses, and Citizen Bartélemy himself, with whom a left a sample of the one fallen at Barbotan in 1790, had never seen anything coming close to it. I found myself thus enlightened on one of the main points of my mission. I similarly examined the collection and the cabinets of the Central School, and if I did not find there anything analogous to the object of my research, I nevertheless brought back with me a most deeply felt esteem for the zeal, the efforts and the perseverance of the professors who compose this establishment.[6]

---

[5] Ecoles Centrales: public establishments of learning founded by the Revolution in replacement of the ancient colleges, with an emphasis on scientific education and financed by the State. They had to be equipped with a public library, a garden, a cabinet of natural history and a cabinet of experimental sciences (i.e. a laboratory).

[6] J.-B. Biot is touching here, in discreet dissent, on a politically sensitive subject: Bonaparte's hatchet had already fallen on the Ecoles Centrales. They were to be disbanded that same year. They had been found to coordinate poorly with primary education, to suffer

Citizen Lamagdeleine, the prefect, not having himself been able to give me any information, furnished me most obligingly with all means to obtain these at L'Aigle and in the various places where I would be stopping. The librarian of the central school, a young man full of talent and energy, agreed, upon my asking, to gather some information relative to the meteorite of L'Aigle. He was able to collect but simple stories, by word of mouth, but these agreed with each other and with what we already knew. Having no other exceptation concerning the purpose of my mission, I left Alençon on 10th Messidor (29 June) et made myself on the way to L'Aigle, with an active and intelligent guide. I intended to stop in all the places where I could hope for answers to my questions; I even planned to go off to the houses I could make out at a distance from the road.

The first inhabited place we came to was Seez, a small town 10 leagues (40km) from L'Aigle. The noise of the meteorite had been heard there; the people indicated precisely the day, the hour and the diverse circumstances. It was like a very loud thunderclap which seemed to come from the north and the rolling of which, accompanied by several successive explosions, lasted five or six minutes. Persons finding themselves in the square believed at first that it was the noise of a carriage rolling over the cobblestones and coming from Argentan or from the town of Merlerault; they only noticed that they were mistaken when nothing was seen coming, and the noise was continuing. These persons were the more astonished because the sky was perfectly serene, without the slightest

---

from the absence of moral and religious education, and from too much freedom being granted to the students. The revolutionary schools were found to be too revolutionary... Through the law of 8th Pluviose of the year 11 (28th January, 1803) their newly assembled libraries had been handed over to the communes.

cloud, and one could notice nothing extraordinary in it. They also said that travellers coming from Falaise and Caen had strongly heard the same explosion, and that they had been much frightened; they also added that a ball of fire had appeared in the direction of Falaise, and that the subprefect of Argentan had been given a stone which had fallen from the sky.

These informations made me think that the effects of the meteorite had covered a space much larger than the one we had imagined. My goal being to first circumbscribe exactly said space, I followed the indications I had just received and headed in the direction of Argentan.

We had been already on the road for a while when we met a man who knew my guide and who appeared to me, like him, very intelligent. This man, when interrogated about the phenomenon which I was tracing, remembered very well the day and the hour. He was busy writing when he heard the explosion. His window being open, and oriented to the north, he had lifted his head in order to make out where the noise came from; but, to his great surprise, he had seen the sky serene and there was nothing in the air. He added that people having come back from Caen had heard the same noise at the same time, but that no stones had fallen in that direction; that the one which had been given to the subprefect of Argentan came from somewhere else, and that the noise had generally seemed to come from the north-west, and to follow in a parallel the road from Argentan to Falaise.

This was precisely the direction indicated by the letters we had received. Upon this information, we turned around and took again the road to L'Aigle, well assured that we were leaving nothing of importance behind.

We stopped at Nonant, a village situated 8 leagues (32km) to the west-south-west of L'Aigle. The inhabitants there have heard very distinctly the explosion of the meteorite. It scared them enormously; they compare it to the noise of a carriage rolling over cobblestones, or to the noise of a violent fire in a chimney. Barrier workers, who had laid themselves to rest along the road, jumped up, much afraid. They saw nothing in the air, which was serene. No stone fell in that place.

From Nonant, we went to the town of Merlerault. On our way, we met shepherds in the countryside. I questioned them, asking them if they had not been much scared by an extraordinary noise which had made itself heard some two months ago. They answered affirmatively, indicated to me exactly the day, the hour and the direction of the noise. They too had been surprised to see a serene sky. Other peasants whom I questioned on the road reported the same.

At the town of Merlerault, seven leagues (28km) west-south-west of L'Aigle, I collected the same stories; but the noise of the explosion and the fear it induced were stronger by reason of the increased proximity. Men, women, children whom I questioned agreed exactly as to the day, the hour and the direction of the meteorite. They had seen nothing in the air, and the sky was serene. Horses which were in a courtyard, having come back from the fields and still harnessed, jumped in fear over a hedge and ran into the street: so great was the strength of the explosion, yet at a distance of over 7 leagues (28km). No stones had fallen in this town; but people had heard that some had fallen in the direction of L'Aigle, and I was given one of these stones which had been brought as a curiosity by a carter. It was indeed a piece similar to those which had been sent to us.

From Merlerault we went to Sainte-Gauburge. On the road, I questioned a great number of peasants, on their way to or at work in the fields. Men, women, children all had heard the explosion the same day and report it to have been at the same time, on a Tuesday, between noon and two o'clock.

A little cauldron-mender of some ten to twelve years of age, who was on the road with his sheets of tin and his tools on his back, was listening to a peasant woman from whom I was asking details about the explosion. Oh! Monsieur, he said to me, it was heard much farther away; it was heard three leagues (12km) from Avranches. - Is that what you have heard said? - Monsieur, I know it better than from hear-say, because I was there. - There are 36 leagues (144km) from Avranches to L'Aigle.

In the village of Sainte-Gauburge, 4 leagues (16km) to the south-south-west of L'Aigle, the inhabitants have all heard the explosion the same day and more or less at the same time than everywhere else; but no meteoritic stones fell at this spot. Yet, they had heard about those which had fallen near L'Aigle, and several inhabitants of the place were in possession of some samples. I was taken to a thatched cottage outside the village, where I found a peasant of the area who held one in his hands. I first showed him the one from Barbotan, and he immediately identified it as one having fallen from the sky. Then he showed me the one in his possession: it was in every way similar to the ones we had, and must have weighed about 0.48 K (1.06 lb). His wife had picked it up in front of their door, where it had fallen and had dug itself into the ground. The stone still showed the traces of this fall and the peasant brought them to my attention. He seemed to be holding onto this curiosity: I did not ask him for it. He told me that he was from the village of Saint-Sommaire. I have since found out that this is the canton where the greatest number of them fell.

An old man told me that, finding himself then working in a field near L'Aigle, he had seen in the sky a small cloud from which explosions came out which succeeded each other over several minutes; he had heard stones whistle and fall.

From Sainte-Gauburge to L'Aigle, I questioned several

peasants who all agreed with the reports I had already gathered. Night falling prevented me from further multiplying these informations, which at any rate could not have taught me much more, as it was to the other side of L'Aigle that the meteorite had burst. I reached this city around ten in the evening, on the same day as I had departed from Alençon.

I went immediately to see our colleague Leblond; but I was not able to see him. I learned anyway that the whole city had heard, at said day and hour, a terrifying noise. No stones had fallen in L'Aigle itself, but people had heard about them. People who were then at Caen assured me that the same noise had been heard there at about the same time, and that a globe of fire had been seen, which had caused a great fear.

On the day following my arrival, I showed up at our colleague Leblond's: I was as happy as I was flattered to find in him both the lights of a scientist and the kindness of a friend.

Citizen Leblond and his brother-in-law Citizen Humphroy, a former military man, had both, as well as their families, heard the noise of the meteorite. It was like a thunderclap which lasted without interruption for about five minutes, and which was accompanied by frequent explosions resembling discharges of musketry. At first, one had taken it to be the noise of a carriage rolling over the cobblestones, or for the noise made by a violent fire in a chimney.

Comparing these accounts, made by enlightened men, with those we had collected in the countryside over an aera of over 10 leagues (40km) of radius, we can see that they absolutely agree as to the day, the hour and the nature of the explosion. We can therefore, with all certitude, deduce from them the following consequences:

There occurred in the vicinity of L'Aigle, on Tuesday 6th Floreal in the year 11, around one o'clock in the afternoon, a violent explosion which lasted five to six minutes, with a continuous rolling. This explosion was heard at a distance of some 30 leagues (120km) all around.

If we compare the account made by the courier of Brest, about the globe of fire which he had spotted, with that of the travellers coming from Caen and from Falaise and with the content of the letters written from this latter city on the very day of the explosion, we shall find that these accounts agree as to the day, the hour and the direction of this meteorite.

I have learned since, through other information, that the phenomenon has been seen nearly at the same time at Pont-Audemer and in the vicinity of Verneuil.

From these testimonies put together, we can additionally deduce as certain this second consequence:

On Tuesday, 6th Floreal of the year 11, a few moments before the explosion at L'Aigle, there appeared in the air a luminous globe animated with a rapid movement. This globe was not observed at L'Aigle, but it was in several other cities of the region and which where very distant from each others.

I took all measures necessary to have precise and multiple intelligence of the various spots where this phenomenon was seen, so as to deduct from it its progression, and to follow, as far as possible, the whole length of its course. But meanwhile, if one considers the day, the hour at which it was observed, the trajectory it took, and the explosion which succeeded to its apparition, we shall conclude from it with the same amount of certitude this third consequence:

The explosion which occurred on 6th Floreal in the vicinity of L'Aigle was the consequence of the apparition of a flaming globe which exploded in the air.

And it must be noticed that these results agree perfectly with the descriptions which have already been made of meteorites to which is attributed the fall of mineral masses.

I come now to the very question of the fall of such masses; and as this was the most important element in the phenomenon, it is the one to which I dedicated the greatest care, both in details and in time.

The first information I received at L'Aigle about this object were given to me by Citizen Humphroy, and relates to a stone weighing 8.56 K (18.87 lb) which is said to have fallen in La Vassolerie, a village situated one league (4km) to the north of L'Aigle. Citizen Humphroy, guided by the public rumor, went to the spot on the very day, on the advice of his brother-in-law, Citizen Leblond. He had still seen the peasants assembled around the hole the stone had made upon falling. The stone was already reduced to 6.1 K (13.45 lb) because everybody hastened to take pieces from it. Citizen Humphroy obtained easily what was left over and brought it to his brother[-in-law], who sent it immediately to Paris. I have in my possession a characteristic sample of it.

Citizen Leblond, aware of the importance of the phenomenon, transported himself immediately on the spot. He too still saw the peasants assembled; he noted with them the depth of the hole, which was 0.5m (19.69 in); he saw the dirt thrown up all around it to more than 4.86m (15.94 ft) distance; He dug from the hole three large silexes which seemed to have prevented the stone from penetrating to a greater depth.

I have since seen in his company this scary trace of the meteorite, I have listened to the accounts of the owners of this house, I have heard the testimony of the children who were staying in the house when the mass fell twenty paces away from them; and these are the informations which I received from them:

The father of these children was returning from L'Aigle with his wife and his daughter-in-law; they suddenly heard in the air an extraordinary noise of thunder, accompanied by a rolling noise as of a big fire in a chimney. There were almost no clouds in the sky, except for one small black cloud, and a scattering of others such as are commonly seen; but there was no sign of a thunderstorm. The noise seemed to come out of the small cloud, and moved away in front of them continuously roaring and puffing. They were all three extremely scared. The young woman fell in a faint and the father did not dare to speak. This frightening noise lasted only a few minutes. When they arrived home, they saw all their neighbors assembled, and thought that some misfortune had occurred during their absence: they drew closer and were shown the mass which had just been unearthed. The father weighed it immediately: it's weight was 8.56 K, as I have reported.

The son, having come back from the fields, gave me details even more precise: it was he and his brothers who had first come running at the noise of the fall of the stone, and who had unearthed it.

He was having lunch with his brothers and sisters under a walnut-tree which he pointed out to me: suddenly, they heard above their heads a terrifying noise of thunder, accompanied with a rolling so continuous that they believed that they were about to die. The yound man told his siblings to throw themselves to the ground so as not to be blown away. Then they heard in the neighboring meadow a terrible blow, which they compare to that of a full barrel falling from on high. They ran to the spot, from which they were separated by a hedge, and saw this stone, which had dug itself in so deeply that it had made the water well up.

I examined with our colleague Leblond the hole from

which this mass was pulled. It is situated at the entrance to a humid pasture, the soil of which certainly does not include anything like it among its natural products. Can one reasonably suppose that a mass so considerable could have existed for a long time without being noticed in a place frequently passed by; that all of a sudden the children of the house and the neighbors would have gotten together, just by chance, to affirm that they had heard something very heavy fall on that same spot, with a very big noise; that all these circumstances would have coincided with what was happening at the same moment two leagues (8km) away from there, and that finally none of the spectators could have remembered ever having seen this stone before? Yet these are the pecularities one would have to suppose having come together in order to infirm the truth of this testimony.

Let us observe another very important circumstance. As the peasants had, on the very spot, and in a few instants, broken off so many fragments of this mineral mass, it appears that it did not then have the extreme hardness which we find it to have today. In fact, our colleague Leblond assures us that when it was brought to his home, it was still very easy to break, and the small pieces which one broke off from it crumbled between one's fingers. This is assuredly a fact attested by an eyewitness worthy of all trust. The same thing has been affirmed to me since in twenty different places, and by all those who handled theses substances in the first moments. Yet, so fast a passage from a friable state to one of complete solidity announces the presence of a cause which had recently troubled their aggregation. This agrees therefore with the testimonies, proving that these mineral masses are foreign to the places where they found themselves, and that they had been recently transported there.

Returning to La Vassolerie, I took pains to acquire

specific informations to acquire a knowledge of the route which the meteorite had followed, and the expanse of land over which it seemed to have burst. These first informations gave me as limits the city of L'Aigle on the one side, on the other five villages, with names of Saint-Antonin, Gloss, Couvain, La Ferté-Fresnel and Gauville. It was an area three leagues (12km) long by two leagues (8km) broad, which I determined to explore completely the following day.

I left at six in the morning, accompanied by a guide who knew the country and inhabitants well. We went first to the château of Fontenil, where all the testimonies placed the beginning of the explosion. The masters being away, I talked to the concierge of the château, who seemed to me a sensible and trustworthy man. He had, like everybody else, heard several violent blows, resembling cannon fire, followed by a roaring like a fire in a chimney. All of a sudden, a huge thud had been heard on the ground of the enclosure surrounding the château, as of a large tree falling after having been stripped of its branches. Workers which where at work in a nearby wood came running at the noise; the cattle, fightened, ran towards the place where the fall had occurred. A young man of fifteen, who was working ten feet away, under a shed, said he had seen a stone fall: they all got closer and retrieved one weighing three pounds. It had made a hole of 18 inches in the ground. The concierge measured it after having removed it carefully, in order to deposit it in the archives of the house together with an account of the event. I saw the young man who is an eyewitness; I also saw the hole made by the stone; I saw the very stone itself, and am bringing back a sample of it, which I was allowed to break off.

The soil of the enclosure, which is called in this area the "cour," is made of loam, humid and covered with grass. Below the humus one finds pebbles: nothing signals that one should find there naturally substances resembling meteoritic masses, and all the inhabitants of the house are quite certain never to have seen any.

I have also a piece of a similar sample, fallen into a field near Fontenil: it passed whistling above the head of the shepherd, to whom it caused a great fright, and fell twenty paces away from him. The sheep, terrified by the noise of the meteorite, huddled together. The field has since been plowed, and no other stones of the same nature have been found. These details were given to me at Fontenil by an eyewitness who was brought before me.

From Fontenil, I went to the hamlet of la Métonnerie, and the concierge of the château which we were leaving was kind enough to accompany us to this farm, which is his property. The inhabitants of the farm saw the cloud above their heads. Their account of the noise of the explosion is the same as everywhere else. They saw two stones fall into their yard, quite close to them: one, of which they showed me the spot, whistled in falling; it was burning hot, as the earth was seen to smoke all around it. They only dared remove it the following day, so much afraid they were. I am bringing back a sample of it. The second one had fallen into a hedge: they looked for it for a longtime, but could not find it.

The soil of la Métonnerie is made of a little humus covering a layer of shale; above, there is gravel which is used in construction.

I have also a sample of a stone fallen nearby, in a place called la Marcelière. It was seen by a child who was guarding sheep; it fell next to him. The piece I am bringing back was given to me by the father of this child. According to the volume he evoked to me, this stone must have weighed about 1.96 K (3 pounds = 4.32 lb) before anything was taken away from it.

From la Métonnerie, I went to the village of Saint-

Nicolas-de-Sommaire: I introduced myself to a lady to whom had been brought many meteoritic stones; she used to hold the seigneury of this canton. She welcomed me in good style and passed on to me, by herself and through her attendants, all the details which had come to her knowledge. I found at her home two priests, the local one and the one of a neighboring hamlet called Saint-Michel-de-Sommaire; there was besides these the forest warden, and a trusted lady, long attached to the house. All these persons, except for the warden, are eye-witnesses of the fall of the stones. The latter was then on his way from L'Aigle; he only saw the meteorite and heard the noise.

The priest of Saint-Nicolas was looking straight into the cloud from which the explosion started. It was an elongated square, the longer side of which was oriented east to west; it seemed immobile, and a continuous noise was coming out of it, resembling the noise of a great many drums; then one could hear the stones whistling through the aire like passing shots, and fall onto the ground with a dull sound. One could very well see that the cloud exploded successively from different sides, and each one of these explosions resembled the noise of a fire-craker. The priest of Saint-Nicolas heard these stones fall, but without seeing them falling; but the priest of Saint-Michel assured me that he had seen one which fell whistling into the courtyard of his presbytery, at the feet of his niece, and it rebounded more than one foot high on the cobble-stones. He right away told his niece to bring it to him; but she didn't dare, and another woman who was nearby picked it up. I did not see it; but the priest assured me that it was in every way similar to the others, and these stones, of which we have a great number before our eyes, are too well known now in this area that it should be possible to mistake them for something else.

The mistress of the house gave me several of these masses which had been seen to fall. I am bringing along some more, of which I was shown the still recent holes, and which bear the imprint of the terrains on which they fell. They are all of the same nature than the ones we already have, and in this regard there are as many witnesses as there are inhabitants. It seems, from the information which I have gathered, that there fell in this place and in its vicinity, a frightening number of stones; but despite the fact that they are still big, as they weigh up to 0.97 K (2 pounds = 2.14 lb), none of them equal those of la Vassolerie and of the vicinity of Fontenil: a circumstance which it is important to notice.

Everybody agrees in saying that the stones smoked on the spots in which they had just fallen. When brought inside the houses, they released an odor of sulfur so disagreeable that they had to be brought outside again. A large chunck which I broke off still yielded to me this odor very strongly, but only in the inside. During the first days, these stones broke very easily; all have since acquired the hardness which is now known to us. These changes in state are as many physical proofs which converge to show that these stones are foreign to the places where they then found themselves, or that they had been transported there recently.

Here, as in la Métonnerie, the soil is made of loam, covering a layer of shale; all the houses are built of pebblestones: never was anything seen resembling a meteoritic stone.

We must remark that the testimonies here acquire greath strength through the status and the moral qualities of the witnesses. We have at first a very respectable lady, who can have no interest in making things up; then two men of the church, who could not, for any motive, have intended to alter the truth, especially in front of persons on whose esteem and trust they depended; finally, there is an aged woman who seems to have been long attached to this household and who, being convinced that this phenomenon is a warning from heaven, would not have dared to misrepresent its circumstances, especially when talking in front of persons she is in the habit of respecting. Finally, the testimony of the forest warden is in itself a guarantee of the truth of all the others; for I knew already that this man had not been present when the stones were falling, and neither did he pretend to have seen them fall. Only, his work compelling him to criss-cross the fields, he had had the occasion to spot and unearth several of these masses, which he gave to me, and of which he showed to me the still recent holes. He was quite certain never to have seen anything similar, and one knows how keenly observant the people of his profession are.

From Saint-Nicolas-de-Sommaire I went, led by this warden, to the hamlet of Bas-Vernet where he lives, and in which it was said that a great number of stones fell. Seeing the desire I had to find one by myself and to dig it up from the ground, he brought me to a little field which he owns, and where he had noticed a hole which he thought had been made by one of these stones: he had waited for the harvest to be done to ascertain himself of it; but however much we looked and dug in this hole, we could not find anything. If it was an annoyance for me to be disappointed in my expectation, it gave me at least a renewed occasion to judge the good faith of my guide.

We then went to a neighboring farm, where we found an elderly woman and two young girls, who declared to us all three that they had seen stones fall and were horribly afraid by them: they were alone then in the house and expected to perish instantly. They showed to me several holes in the enclosure of the farm, from which they had extracted stones, and gave me a sample. It is always of the same kind.

We tried for a long time to find by ourselves some left-over; but it was in vain. The ground had since been soaked by the rain, the grass had grown, and the holes themselves, from which the stones had been extracted, had refilled themselves almost entirely. It was therefore very difficult to find more, which would have escaped the first searches. We looked especially under a tree and in a hedge where someone had heard stones fall between the branches, and from which a bird had flown up; but we found nothing. I did nevertheles notice that several of the branches of the tree and of the hedge, situated in a vertical direction, had suffered damage.

After all these fruitless searches, we went to a neighboring farm. We were again given the same accounts about the explosion and the fall of the meteorite. The son of the household, aged ten to twelve, his mother and his sister, aged fifteen or sixteen, were witnesses to the facts. In the middle of this terrifying noise, which they describe like everybody else, they saw a big stone fall, which broke off a branch of a pear tree: the young lad ran to pick it up; but finding it dug into the ground, he called onto his sister to bring a spade. She came running; but hardly had she arrived that a small stone passed in front of her face and fell at her feet. She then had no greater hurry than to flee, and the stone was picked up the following day, when fear had dissipated, along with the danger. I was shown the pear-tree, and I am bringing a sample of the stone which broke one of its branches.

Several other farms in the vicinity gave me the same testimonies and everywhere the same phenomena have been seen.

I left those parts to go to the hamlet of Mesle, to a farmer named Gibon, whom my guides knew. He was a man of sixty-four, full of good sense and reason; he received me with the greatest cordiality. He, his family and his servants, are eye-witnesses to the phenomenon; they describe its circumstances exactly as they are everywhere else. The roaring ressembled so much the noise of a chimney-fire that they believed that the house was on fire, and ran to the pond to fetch water to extinguish it. "We saw, this old man told me, stones fall from above. I myself, who am not fearful and who was tired, did not bother to go pick them up; but my children ran out for them and brought them back. One of them fell close to the pond, and scared a hen which was close by; another fell onto the top of our roof and rolled all the way down: we thought that the chimney was falling." Looking at this respectable farmer, on could not doubt that his testimony was the exact expression of the truth.

I was given a sample of this stone; I was shown on the slope of the roof the wooden link which serves to hold the thatching and which it had broken off. A lot of other stones had fallen into the courtyard which had been picked up. They assured me that there was one in the pond, and another one in a half-dried up ditch. We had to give up on the former; we looked for the second, but to no use.

The son of the household, who had already given to me all that were left over, told me that he had found some in a field, about a quarter of a league (1km) away. I asked him if he had similarly visited all the neighboring fields. He answered me that he hadn't; and as the place he was indicating was close to Saint-Nicolas-de-Sommaire, where I knew a great number of these stones to have fallen, I decided to take up again my search, hoping that I would be more lucky.

Indeed, after looking for about an hour, under a most ardent sun, we found one which I pulled myself from the ground where it was buried; I held it for a long time, burning, in my hand, so great was the heat to which it was exposed. It perfectly resembles all those we already had.

Satisfied of this little discovery, I examined the nature

of the soil where we were, and of the various substances which were to be found in it. I gave to this examination a time and care proportionate to its importance. It is a fairly light soil, on the surface of which one finds gravel and some foundry slag which is called *laitier*. It is said that a very long time ago, there had been on this spot smithies, which had been abandoned. Besides, we know how this slag differs from meteoritic stone, and the peasants themselves did not err about it; for, around L'Aigle, they know nowadays these stones very well, and know very well how to distinguish them from the others, which they call in opposition "natural stones."

Coming back, my young guide pointed at a shepherd in the fields who was known in former times to be an unbeliever, but who had been converted by fear of this terrible meteorite.

Returning to the village of Mesles, I left immediately for the town of Gloss. It was one of those which my previous informations indicated as being on the edges of the meteorite. Indeed, no stones had fallen there, despite the fact that the explosion had been heard violently to the south-west. I knew that a few stones had fallen, but small ones and in a very small number, at the hamlet of la Bélangère, situated to the west of Gloss. Through these accounts, and through the informations I received about it, I found myself confirmed in my opinion that nothing had fallen in the villages of Saint-Antonin and of Couvain.

From the roaming around I had just done and from the informations it had yielded me, I knew the limits of the explosion to the south, the east and the north; it remained for me only to wander along the western edge, and therefore, when I left Gloss, I directed myself towards the south-west.

I went first to the hamlet of la Barne, into the house that bears this name. The persons living in it had heard the noise of the meteorite, and had been much scared by it; but having found themselves inside their home, they had seen no stones falling, and were only alerted about this phenomenon by their tenants who brought pieces which had just been found in the courtyard. I was given a sample.

The master of the house took me to his farm, the tenants of which furnished me with much stronger testimony. Not only had they seen and heard the meteorite, but stones fell whistling around them like hail. They ran to the pond, believing that the buildings were on fire; their fright was such that they expected to die and they still talked about this event only with terror. All the stones fallen here are very small: these people had picked up so many of them that they ended up throwing them into the barnyard, as being of no interest. Still, I was given a few of them, which they had preserved. We looked for a long time among the pastures to see if we could still find some on the ground; but in vain: the grass had grown too high. They don't say here that the stones were hot when they were picked up; this is probably due to their small volume.

From there I went to the hamlet of Boislaville, and introduced myself at the home bearing this name. The owner, whom I found myself addressing, is a young man of twenty-eight to thirty years, who seemed educated and well-born; he served in the war of the Revolution and is therefore not likely to be afraid by a thunderclap. These particularities giving much weight to his testimony, I took it down with special care, and I report about it faithfully.

Citizen Boislaville was standing in the middle of his courtyard, bareheaded; he suddenly heard something like three or four cannon shots, followed by some kind of a discharge, which sounded like a fusillade, after which there occurred like a horrendous roll of drums, accompanied by the very loud whistle caused by stones falling to the ground. The air was quiet and the sky serene; one could only observe directly above the courtyard a small black cloud which seemed immobile and from which all this noise seemed to originate. Right afterwards, a great quantity of meteoritic stones were picked up in the enclosure around the house: they were all extremely small. Citizen Boislaville gave me several pieces.

The mother of Citizen Boislaville, an aged and very respectable lady, attested to the same thing and with the same details. All her help had seen the same effects, and their accounts agreed among them. They had been extremely scared; the animals were violently agitated and all believed that the fire was everywhere in the house.

Citizen Boislaville had informed himself whether stones had fallen in the town of La Ferté-Frénel; but none had been seen, and this matched the accounts which had been given to me in other places.

Here, as in la Barne, the soil is made of good loam, as are the fields and pastures around; one cannot find pebbles, and they build with bricks. Citizen Boislaville is quite certain that one has never seen in this countryside stones similar to those which fell.

Here is therefore a witness whose moral character puts him above suspicion of untruth. His account coincides in the smallest details with those reported everywhere around. Such an account could it exist, if it were not based on truth?

From Boislaville, I went to the farm of la Blandinière, where I had been told that meteoritic stones had fallen in a fairly great quantity, but they were very small. I only found in the house an aged woman who could not give me much details, but who confirmed what I already knew. From there, I came to the hamlet of Teil, where I expected to find very few of these stones; indeed, only a small number had fallen there, and they were very small. It was therefore difficult to get samples, the inhabitants holding onto to them, the more so because they were rare. I experienced the same difficulty, for the same cause, in the neighboring village of Guillemins; yet I was given one of these stones which had fallen in front of the door of a house together with several others which I was shown, and which were similarly of a very small volume. From all these signs, I judged myself to be at the western limit of the explosion. Indeed, I assured myself by pushing farther that no meteoritic stones have been seen beyond this place; none fell in the town of Gauville.

Taking up the road again towards L'Aigle, I stopped at the château of Corboyer. I knew that many stones had fallen in this place. Indeed, the workmen who were working in the courtyard told me that they had been much afraid when they had heard them whistle around them, and seen them come down the roofs, like hailstones. The owner was absent; I talked to the concierge, who seemed to me a very intelligent man. He confirmed all these facts and took me to the mayor of the place, who gave me a piece which had fallen in front of his house and assured me that none like it had ever been seen in the area.

On the day following the explosion, the mayor had written to the subprefect of Argentan to inform him of this appalling fall of stones; he had even sent along a sample, and it was the one which I had been told about at Seez. But, before writing to Alençon, the subprefect had deemed necessary to get further informations, which were held up by various circumstances. This was why Citizen Lamagdelaine had had no knowledge of the event.

I returned to L'Aigle at ten in the evening, bringing with me all the samples I had been given, as well as the accompanying notes, which I had taken down on the spot; the following day, I busied myself putting them in order. Despite the fact that these informations seemed to me sufficient to establish the reality of the phenomenon, I neglected nothing during my stay in L'Aigle in order to complete them, and I looked with equal good faith for anything that would confirm or infirm them; but, as to the latter, I could not find any plausible objection, and especially, not a single observation, not a single story told on the spot which would have contradicted the results of my enquiry.

Still, I wanted to employ yet another, last means to verify it. It is a usage among the peasants of the area to assemble on Sunday mornings on the square at L'Aigle. On that day, I went among them, asked them questions and, according to the accounts they made of the meteorite, I was able to consistently determine in which canton they lived; for those who had seen stones fall were within the limits I had roamed, and those who hadn't seen any, were outside. There occurred no exception to this rule. I concluded from it that I had well circumscribed the surface over which the meteorite had burst.

It was amidst these groups, where nobody was surprised to see so much importance being given to the phenomenon, that was pointed out to me the one among all the peasants of the area who seemed to have been exposed to the greatest danger. He is called Piche, a wire stretcher, who lives in the village of Aunées, in the commune of Gloss. When the explosion happened, he was working outdoors with several workmen: a stone glanced along his arm and fell at his feet; he tried to pick it up but it was scorchingly hot and he let it drop, quite scared. This fact, which was told to me at first on the square by the peasants, was confirmed to me by this man himself, when they brought him to me. He no longer had this stone, which an interest foreign to the sciences had made it to be picked up and mixed with several others; but he gave me a piece which had fallen at the same time, on the same spot, close to him, under the eyes of all his companions.

Finally, when I had assured myself by all possible means

that I had no more enlightenment to acquire nor any new informations to hope for, I left L'Aigle on 16th Messidor (5 July) and returned to Paris.

If one is to compare, following the rules of critical enquiry, the moral and physical testimonies which I have thus truthfully reported, one will find in them a reunion of proofs such as can fit only the truth.

Indeed, let us consider first the physical testimonies.

Nobody has ever seen, before the explosion of 6th Floreal, meteoritic stones in the hands of the inhabitants of the area.

Mineralogical collections made with the greatest of care, over several years, in order to gather the products of the département, do not include anything similar; the memoirs held by the council of mines about the mineralogy and the geology of the area of L'Aigle make no mention of it.

The founderies, factories, mines of the area which I have visited have nothing in their production and in their slag which has the slightest connexion with these substances. One cannot find in the land any trace of a volcano.

All of a sudden, and precisely since the time of the meteorite, these stones are found on the ground and in the hands of the inhabitants of the area, who know them better than they know any others; they are so common that one can estimate the number of those which are shown around to be two or three thousand.

These stones are encountered only in a well determined area, on lands which are foreign to the substances which make them up, in places where they could impossibly, given their sizes and their number, have escaped attention.

The biggest of the stones, when broken open, still release a very strong sulfurous odor in their inside; this has disappeared from their surface and the smallest of them no longer release any that can be made out: in such a way that the odor released by the bigger ones seems of a nature such that it must also disappear with time.

All these are as many physical proofs attesting to the fact that the meteoritic stones of the area around L'Aigle are foreign to the places where they have been found; that they have been transported there recently, since the time of the explosion, and by a cause which has modified the principles which make them up.

If now we consult the moral testimonies, what do we find? Twenty hamlets scattered over a surface of over two square leagues (64 sqkm), almost all the inhabitants of which come forward as eye-witnesses and testify to the fact that a harrowing rain of stones was cast by the meteorite. In their numbers one finds grown men, women, children, old people; they are simple and rough peasants, living at big distances from each other; farmers full of good sense and reason; respectable men of the church, young men who, having been to the military, are immune against fear created by delusions: all these people, of such different professions, mores, and opinions, who have only little or no relations to each other, are all at once ready to agree upon the same fact which they have no interest in making up; they all report it to have happened on the same day, at the same hour, in the same moment, in the same circumstances, and using the same comparisons; and this fact, so universally, so strongly attested to, is but a consequence of the physical proofs previously assembled, namely that there fell onto this area stones of a peculiar nature following the explosion of 6th Floreal.

Moreover, there are still being shown traces of debris

which testify materially to the fall of these masses, about which one speaks only with fear. People say that they have seen them come down along roofs, break branches of trees, rebound on the pavement; it is said that the soil was seen to smoke around the biggest of them, and that one could hold them scorching in one's hands. These accounts are made, these traces are shown only in a delimited area of land. Only there is it still possible to find some meteoritic stones; there is not even one which is known to have been found on the ground outside of this area, and there is not one single witness who maintains to have seen stones fall anywhere else.

Finally a third category of proof results from certain physical particularities unanimously reported by the inhabitants of the area, who are too little educated to have foreseen their consequences: I mean to say the successive changes observed concerning the hardness of these stones and the odor they released; which changes, according to the account of the witnesses, among whom must be counted our colleague Leblond, have intervened in the space of a few days following the explosion of the meteorite; changes the traces of which I have myself observed very clearly by breaking off pieces of different dimensions; and this new comparison between testimonies and facts only serves to show yet more agreement between them.

In this way, all proofs, be they physical, be they moral, which it was possible to collect, concentrate and so to speak converge towards a single point; and if one considers the way in which we have been lead, by comparing testimonies, to the spot of the explosion, the number of informations taken on the spot, and the fact that they matched those which were collected at ten leagues' (40km) distance; the great number of witnesses, their moral character, the similarity and perfect coincidence of their accounts, wherever they came from, without there having been discovered even one single exception, we shall conclude without the slightest doubt that the fact upon which all these proofs unite has really happened, and that *stones fell in the vicinity of L'Aigle on 6th Floreal of the year 11*.

Then, the full set of testimonies shall yield the following description of this phenomenon:

On Tuesday, 6th Floreal of the year 11, around one o'clock in the afternoon, by serene weather, was seen from Caen, Pont-Audemer and from the vicinity of Alençon, from Falaise and from Verneuil, a flaming globe, of a very brilliant radiance, and which moved through the atmosphere with much speed.

Some moments later there was heard at L'Aigle and around this city, in an area of a radius of over thirty leagues (120km), a violent explosion which lasted five or six minutes.

There were at first three of four blows resembling cannon shots, followed by some kind of discharge resembling a fusillade; after which one heard something like a terrible rolling of drums. The air was quiet and the sky serene, with the exception of a few clouds, such as are seen commonly.

This noise came out of a small cloud which had the shape of a rectangle, the long side of which was directed east-west. It seemed to be immobile during the whole duration of the phenomenon; only the vapors of which it was composed shot out momentarily from various sides under the effects of the successive explosions. This cloud was more or less half-a-league (2km) to the north-north-west of the city of L'Aigle: it was very high up in the atmosphere; for the inhabitants of laVassolerie and of Boislaville, hamlets which are distant of more than one league (4km) from each other, observed it at the same time above their heads. In the whole canton over which the cloud hovered one could hear whistlings similar to the one of a stone thrown by means of a sling, and in the same time one saw a great number of solid masses fall which are exactly similar to those which have been designated under the name of meteoritic stones.

The area upon which these stones have been thrown has as its limits the château of Fontenil, the hamlet of la Vassolerie and the villages of Saint-Pierre-de-Sommaire, Gloss, Couvain, Gauville and Saint-Michel-de-Sommaire.

It is an elliptical area about two leagues and a half (10km) long by about one league (4km) wide, the longest dimension being directed from south-east to north-west, with an angle of ca 22°: this is the present direction of the magnetic meridian at L'Aigle.

One can draw from this some informations about the direction of the meteorite. Indeed, if it had exploded in a single instant, the stones would have been scattered over a surface more or less circular; but the duration of the noise signals a succession of explosions which must have scattered the stones along an elongated surface following the direction in which the meteorite was moving. This elongation indicates therefore the horizontal direction of the meteorite; and by matching this result with the testimonies which have the fiery globe falling to the north, one will conclude, with a great appearance of certitude, that the meteorite was going from south-east to north-west, at an angle of ca 22°.

If observations made about the duration of the noise could be considered to be exact, one would deduce the horizontal speed of the meteorite from the ellipticity of the area in which the stones have been projected; but I am not aware that any precise observation has been made about this point, and in this regard one can base oneself only on the exactitude of instruments, because astonishment will always induce one to increase the duration of a phenomenon which causes surprise through its continuity. One can only presume from this data that the horizontal speed of the meteorite at the time when it exploded was not very high, and it is probably for this reason that it was believed to be completely immobile. This does not prevent it in any way from having had a very great speed in the vertical direction, as the horizontal speed is the only one which observations of this kind can apprehend.

The biggest stones fell at the south-east extremity of the major axis of the ellipse, to the side of Fontenil and la Vassolerie; the smallest fell at the opposite extremity, and the middle-sized ones between these two points. According to the considerations previously mentioned, the bigger ones seem to have fallen first.

The biggest of all of those which have been found weighed 8.56 K (18.86 lb) at the moment when it fell; the smallest which I saw and which I brought back with me weighs only 7 or 8 grams; this latter one is therefore about one thousand times smaller than the first one. The number of all those fallen can be evaluated to two or three thousand.

The samples of meteoritic stones which are mentioned in this memoir have been deposited at the Museum of Natural History. Citizen Thénard has been so kind as to analyze some of them, and he found this:

Silicium:	46
Oxydized iron:	45
Magnesium:	10
Nickel:	2
Sulfur:	ca 5
	108

From which must be removed the quantity of oxygen which has bonded with the metal during the operation. The various pieces which have been tested comparatively have not yielded appreciable differences, despite having been chosen among those which, by virtue of their aspect or by the locus of their fall should have distinguished themselves maximally one from the others.

One can see, from this analysis, that the stones fallen in the vicinity of L'Aigle are constituted following the same principles as the meteoritic masses known up to now; they merely contain a little less magnesium and a little more iron.

These results are quite in agreement avec those which Citizen Vauquelin had already obtained when analyzing the first samples sent from L'Aigle to Citizen Fourcroy.

Besides, whatever the origin of these stones, one must not be surprised to find some differences in the relationships between the substances which compose them, as they are united by mere aggregation, not through intimate combination.

I have limited myself in this relation to a mere exposition of the facts; I have set myself the task to see them as anybody else would have seen them, and I have put all my care in presenting them with exactitude. I shall leave to the sagacity of the physicists the numerous consequences that can be drawn from them, and I will estimate myself fortunate if they find that I have succeeded in removing beyond the reach of all doubt one of the most astonishing phenomena that men have ever observed

Fragment of the météorite of l'Aigle (Musée Vert of Le Mans)

L. Martel, www.psrd.hawaii.edu

Fragments of the l'Aigle meteorite at the Museum of Natural History in Vienna

Jean-Baptiste Biot (1774-1862)

Jean-Baptiste Biot, physicist, astronomer and mathematician, was born April 21, 1774 in Paris. His father, who descended from a peasant family of Lorraine, worked for the Treasury. While studying at the prestigious College Louis-le-Grand in Paris, Jean-Baptiste drew the attention of the mathematician Antoine-René Mauduit (1731-1815) who gave him private lessons. Yet his father sent him to Le Havre into apprenticeship with a merchant. His work consisted mainly of copying letters. He ran away and, at age 18, he joined the Army of the Revolution as a volunteer cannoneer and was injured at the Battle of Hondschoote, which liberated Dunkirk from the armies of the First Coalition of European powers attacking the Revolution. He returned to his family with a permanently damaged knee, got involved in political protests, even marched against the Convention on 13 Vendémiaire, and was accused of being a deserteur and jailed. The mathematician Gaspard Monge, the inventor of descriptive geometry and an important Jacobine, got him out of trouble.

He started to study higher mathematics, entering the engineering school of the Ponts et Chaussées ("bridges and roads"), and became an engineer of public works. For four years, he taught mathematics at Beauvais. He became close friends with the mathematician and engineer Barnabé Brisson (1777-1828), who would become one of the foremost engineers and canal-builders in France, and whose sister he married in 1797 (Brisson himself married a niece of Gaspard Monge.) A protégé of the astronomer and mathematician Laplace, Biot became titulary of the chair of mathematical physics at the College de France in 1800, at age 26.

In 1803, Biot investigated the fall of the L'Aigle météorite, and demonstrated in his report the exoterrestrial origin of meteorites, which made him famous all over Europe.

The following year, 1804, he undertook a perilous ascension in a hydrogen balloon with the chemist Louis Joseph Gay-Lussac (1778-1750) in order to study the electric, magnetic and chemical properties of the atmosphere, and especially to determine the inclination of the Earth's magnetic field. Until 1826, he shared with Gay-Lussac the teaching of the course of physics at the Faculté des Sciences in Paris. In 1809, he became the titulary of the chair of astronomy there.

Biot and Gay-Lussac's ascension in a hydrogen balloon (late XIX century lithograhy).

In 1812, Biot turned his attention to the study of optics, particularly the optical properties of mica, and studied the polarization of light. His work in chromatic polarization and rotary polarization greatly advanced the field of optics and has led to many breakthroughs, such as liquid crystal displays (LCDs), [7] which are used on television and computer screens, and polarizing filters, used in photography.[8]

Always a "hands-on" scientist, he undertook several expeditions for geodetic measurements in France, Spain, Scotland and Illyria, a.o. prolonging the measurements of the Paris Meridian (then the primary meridian) to the Shetland Islands and the Balearics. Armed with a Borda-pendulum, he studied gravimetry (measuring the strength of the gravitational field) at the 45th parallel in Bordeaux.

In 1820, together with Francesco Carlini, he determined for the first time the vertical deflection of the gravity field through measurements at the top of Mont Cenis, in the French Alps.

Also in 1820, together with the physicist Felix Savart, he formulated the law of Biot-Savart, an equation that describes the magnetic field generated by an electric current, relating it to the magnitude, direction, length and proximity of the electric current.

Biot considered that written expression was an indispensible companion to scientific reflexion and was a fecund writer. He was interested in the sciences of Antiquity, particularly Egyptian astronomy, and wrote, among others, a memoir about the Zodiac of Denderah.

His son Edouard (1803-1850) translated the British mathematician, Charles Babbage (the inventor of the concept of computer programming) into French and made his mark both in railroad engineering (he wrote a *Manual of Railway Construction,* 1834) and as a sinologist. He translated the 11th century B.C. *Zhou Li,*, the "Rites of Zhou," of the *Book of Rites.* He died prematurely, at 46,

---

[7] LCDs use light that is polarized by a filter as it enters the liquid crystal, to allow the liquid crystal to modulate the intensity of the transmitted light. This happens as the liquid crystal's polarization varies in response to an electric control signal applied across it. (Wikipedia, eng. - Jean-Baptiste Biot).

[8] Polarizing filters are also extensively used in photography to cut out unwanted rflections or to enhance reflection. (Wikipedia, eng. - Jean-Baptiste Biot)

possibly of grief after the death of his wife, and his father brought the work to completion and supervised its publication. It remains the translation of reference.

Few scientists were as much honored during their life time as Jean-Baptiste Biot.The name "biotite" was given to a certain kind of mica in honor of his work in optics. The Biot number, a dimensionless number (abbreviated Bi) is used in thermodynamics in heat transfer calculations. A now disused unit of measure of the electric current, the Biot, was also named after him (1 Biot = 10 Ampères).

The year of his death, in 1862, at the age of 88, he published a book of studies on Indian and Chinese astronomy.

A.-M. de G.

Library of Congress

TABLE OF CONTENTS

This translation of J.B. Biot's *Relation of a Voyage to the Département of Orne...* was first published in Q-Mag.org in December 2012

See links:

http://www.q-mag.org/
http://www.grazian-archive.com
http://www.metronpublications.com

Made in the USA
Monee, IL
05 November 2026

67383550R10040